Get to Know Big Cats

JAGUARS

I0815074

By Bray Jacobson

Please visit our website, www.garethstevens.com. For a free color catalog of all our high-quality books, call toll free 1-800-542-2595 or fax 1-877-542-2596.

Library of Congress Cataloging-in-Publication Data
Names: Jacobson, Bray, author.
Title: Jaguars / Bray Jacobson.
Description: Buffalo, New York : Gareth Stevens Publishing, [2024] | Series: Get to know big cats | Includes index. | Audience: Grades K-1
Identifiers: LCCN 2022045084 (print) | LCCN 2022045085 (ebook) | ISBN 9781538286029 (library binding) | ISBN 9781538286012 (paperback) | ISBN 9781538286036 (ebook)
Subjects: LCSH: Jaguar–Juvenile literature.
Classification: LCC QL737.C23 J337 2024 (print) | LCC QL737.C23 (ebook) | DDC 599.75/5–dc23/eng/20220920
LC record available at https://lccn.loc.gov/2022045084
LC ebook record available at https://lccn.loc.gov/2022045085

First Edition

Published in 2024 by
Gareth Stevens Publishing
2544 Clinton Street
Buffalo, NY 14224

Copyright ©2024 Gareth Stevens Publishing

Editor: Kristen Nelson
Designer: Leslie Taylor

Photo credits: Cover, p. 1 reisegraf.ch/Shutterstock.com; p. 5 Giedriius/Shutterstock.com; p. 7 Jo Reason/Shutterstock.com; pp. 9, 24 (fur) worldswildlifewonders/Shutterstock.com; p. 11 Sergey Uryadnikov/Shutterstock.com; p. 13 Vladislav T. Jirousek/Shutterstock.com; pp. 15, 24 (rainforest) Guillermo Ossa/Shutterstock.com; p. 17 Walter Mario Stein/Shutterstock.com; p. 19 Gurkan Ozturk/Shutterstock.com; p. 21 belizar/Shutterstock.com; pp. 23, 24 (cub) Kris Wiktor/Shutterstock.com.

All rights reserved. No part of this book may be reproduced in any form without permission in writing from the publisher, except by a reviewer.

Printed in the United States of America

CPSIA compliance information: Batch #CSGS24: For further information contact Gareth Stevens at 1-800-542-2595.

Contents

Jaguars are big cats!

They have fur.
It can be tan.
It can be orange.

Their fur has spots.
The spots look
like roses.

Spots help them hide.

Some jaguars look
all black.

Most live in South
America.
Many live
in the rain forest.

They like to
live near water.
They are good
swimmers!

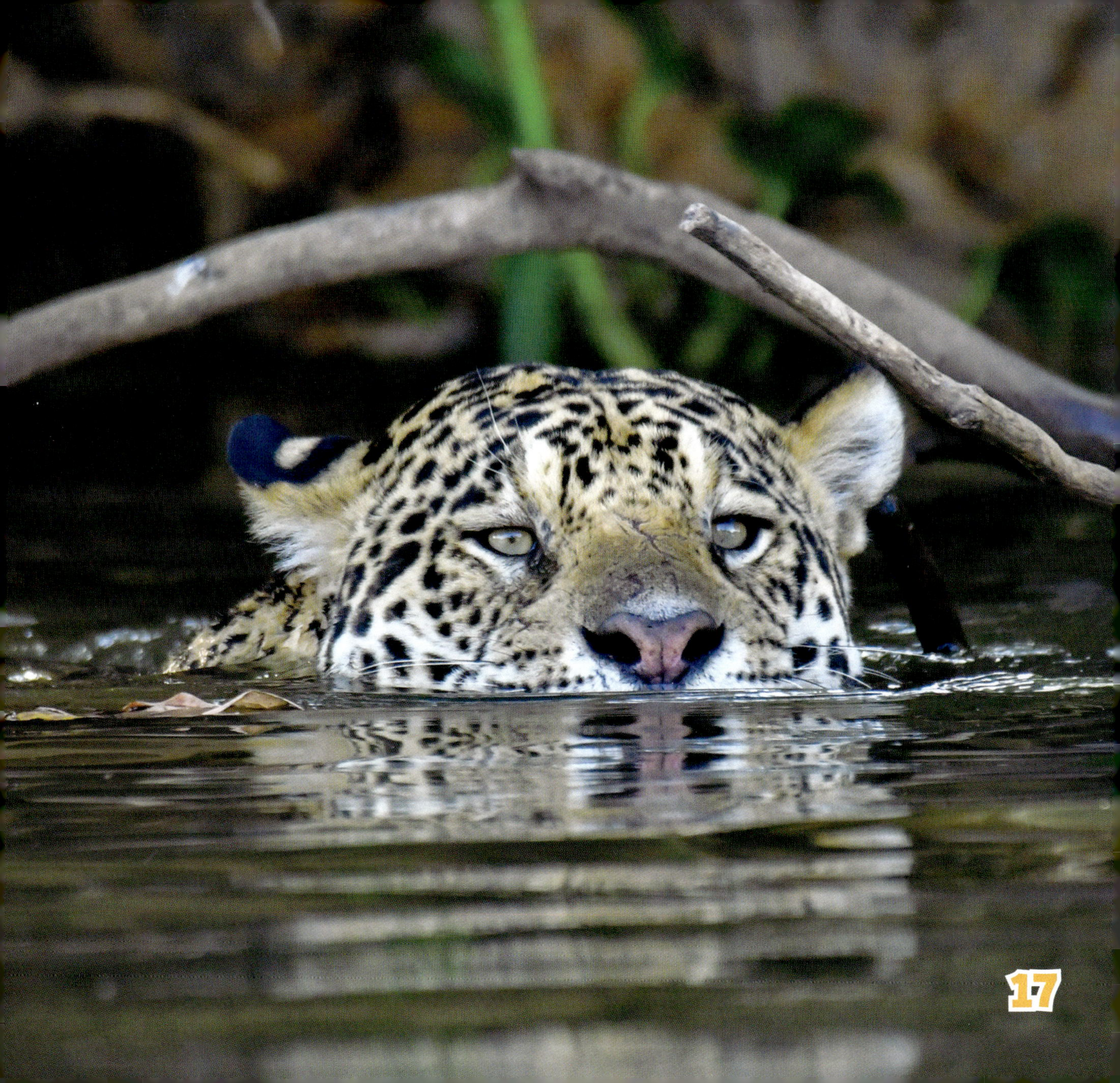

They hunt many animals.
They eat fish.
They eat monkeys.

Jaguar babies are cubs.
They have no teeth.

Cubs grow fast!

Words to Know

cub

fur

rain forest

Index